Evolution

WORK IN PROGRESS

Bruce Alpine

EVOLUTION

WORK IN PROGRESS

Bruce Alpine

Copyright © 2016 Bruce H. Alpine

All rights reserved. No part of this publication may be reproduced, distributed or transmitted in any form or by any means, including photocopying, recording, or other electronic or mechanical methods, without the prior written permission of the Author or Publisher, except in the case of brief quotations embodied in critical reviews and certain other noncommercial uses permitted by copyright law. For permission requests, write to the publisher, addressed "Attention: Permissions Coordinator," at the address below.

queries@brucealpine.com

Contact the Author:
bruce-a@brucealpine.com

Visit The Website:
https://brucealpine.com

Book Layout ©2016 Bruce Alpine

Evolution: Work In Progress

ePub ISBN: 978-0-9941053-7-0
Print ISBN: 978-1-5398826-3-3

Contents

Introduction……………………………….5

Chapter 1… What Has Been………………..10

Chapter 2… What Is Now………………..17

Chapter 3… What Is To Come……………...26

Chapter 4… Are We Still Evolving………....34

Chapter 5… What Is The Future For Evolution.………………………………………….44

Chapter 6… Conclusion……………….. 60

Appreciation………………………………...63

***Intelligence**: The Ability To Adapt To change.*

—Stephen Hawking

Introduction

The main driving force behind evolution is Natural Selection. Charles Darwin was the first to propose the process of natural selection through his most famous publication, *On The Origin Of Species: By Way Of Natural Selection*. Darwin was not the first to realise the importance of this natural process of which, organism that are better adapted to their environments will survive into adulthood and pass on their genes through to their offspring. Adaptations to environments, that benefit, future generations. Alfred Russel Wallace, another naturalist was also working on the same theory in the Amazon River basin, then Malaysia, while Darwin was writing his publication.

After twenty years of writing, Darwin was struggling with the idea of publishing his theory of evolution, as he was fully aware of the ramifications of his findings with the popular religious doctrine of his time.

During a visit back to the United Kingdom and spending time in London to sell some of his specimens he had earlier sent back to Britain. Wallace heard of Charles Darwin and his evolution theory. He contacted Darwin. They spoke intensively together, comparing each others theories. Alfred Russel Wallace informed Darwin of his intention to publish his findings and theory, encouraging Darwin to publish his theory in November 1859.

Upon its first publication, *On The Origin Of Species* sold out on the first day of release to rave reviews. It presented a body of evidence that the diversity of life arose from common ancestry through a branching pattern of evolution. Included, was evidence he had gathered on his five year expedition to the southern hemisphere onboard HMS Beagle, during the 1830s and subsequent findings from further research, correspondence and experimentation.

Darwin was not the first to realise the process of natural selection. But, he was the first to publish his theory, which would go on to become the worlds authoritative scientific explanation and understanding on the processes of evolution and how species have adapted to changing environments, throughout the Earths geological and biological history.

At the time he wrote and published *On The Origin of Species*, and his later best selling publication - *The Descent Of Man*, Darwin would not have known his books would become the Standard model, on how all life has diversified over time. Originating from common ancestors.

Since 1859, findings in many scientific fields have confirmed and indeed greatly expanded on his original theory. We have learned the Earth is old enough for known species to have evolved from common ancestors. The discovery of DNA confirms all organisms are related to one another. And, millions of fossils have been uncovered that provide evidence on how life forms have evolved into another over time.

Over time, many societies have questioned and attempted to express through a diversity of tales on how humans and other life forms have come into being, expressing the universal curiosity these societies have about our origins. Many have been religious and mythology based, in an attempt to understand the world around them. Science is providing evidence, based on findings discovered during scientific methods that are quite distinct from these stories.

Modern humans evolved in Africa over two hundred thousand years ago, over the last sixty-five thousand years, they have migrated around the globe, continuing to evolve to suit their new environments. Their new adaptations benefiting their offspring. Helping to build their communities and cultural beliefs.

Through the course of our ancestors migration, various superficial changes were occurring, creating their distinctive variations and changes from each other. Such as skin pigmentation or colouration between regional environmental differences. Communities living closer to equatorial regions now have darker skin pigmentations compared to communities farther from equatorial regions who now have lighter skin pigmentations.

Farming practices have made dramatic changes, improving food availability and stopping the need to gather and hunt for available food. Farming is most likely to have stopped the need to migrate further, allowing communities to remain in their own unique regions. Eliminating the need for a nomadic lifestyle.

Human evolutionary changes since migrating out of Africa, are many and diverse. Helping humans to adapt and form communities everywhere imaginable on Earth. The changes and adaptations that help humans to survive in tropical areas, extreme cold areas and areas in high altitudes.

Are we still evolving now? That is a question many scientists are trying to answer, especially since publication of Charles Darwin's book, *On The Origin Of Species: By Way Of Natural Selection*. Natural selection has certainly opened our eyes on how we have evolved, and the method of our evolutionary history. Our evolutionary history is documented in our genes. Our DNA is a treasure trove of information into who we are.

Going with the evolutionary method, natural selection, with our evolutionary history and the part natural selection played in that, it is quite reasonable and completely natural to believe we are still evolving and will continue to in the future. The planet is still continuously changing. As man made triggers and natural triggers continue to change our world, The human species will be required to change, and adapt to the planets alterations.

9   Evolution: Work In Progress

Natural selection dictates, species must adapt to changing environments, or become extinct. Species that are not properly adapted will not survive in their environments. Improperly adapted species will not breed normally, to ensure the survival of their species.

Evolutionary changes are still occurring. Even if those changes are superficial, they are still adaptations to our unique environments.

What will the face of evolution look like in the future? In what ways are we going to change in the future, as we continue to evolve? If we are indeed, still evolving.

Chapter one

WHAT HAS BEEN

Since life began on Earth three point seven billion years ago, the environments on Earth have been constantly changing. With the constant changes, a force known as evolution has been constantly changing DNA and protein molecules of every life form and creature through the different ages and eons. Leading up to modern man and the varying species and forms of life that is seen globally on the modern Earth.

A series of events occurred that three point seven billion years ago that resulted in the first life forms emerging, and progressing through to modern times. Events that are ideal for the beginning of life, occuring in just the right sequence and timing. Events and occurances that are just too decisive to be pure chance. Events and occurances that do not require a creator or god to trigger or intiate.

11 Evolution: Work In Progress

Nature was and still is quite capable of performing such marvelous tricks, as a magician performs on stage in front of the live audience. A live audience was not present at the time the first organism emerged in that primordial soup existing on Earth three point seven billion years ago. But, nature began a process that has steadfastly continued and advanced. Various instances presented itself where evolution had to take serious actions and lead a different path, due to serious events, presenting itself that adversely affected the environment, thus affecting the direction those changes were heading.

Catastrophic events occurred where earlier life was seriously compromised. Evolution had to direct itself onto a different path to overcome some serious issues with the environment, a maturing planet presented itself. Subduction zones continuously change the landscape. The old is renewed and replaced. No one can claim recycling is not a natural process. Landmasses unite and divide. Some life forms emerge and thrive in conditions present on Earth. But, as those conditions change, those forms of life that thrived are now compromised and must change as their environment changes, or die out. As a result of Earths continuous changes and various climatic conditions, or re-arranging, over ninety-nine percent of all life that have previously existed on the planet are now extinct.

Evolution is the natural law that ensures life will survive changing environments. As pointed out, some species may not adapt or change accordingly to that change, they will die out and become extinct. But, life in general will survive, ensuring the continuation of other species, where one species will rise to dominate the others.

The first recorded belief in evolution, comes from the Roman philosopher, Lucretius who lived between ninety-nine BC and fifty-five BC. He believed all species alive on Earth were related. Lucretius also believed as times change, nature enforced change. Transformed species in order to adapt to natural changes to their environment.

Since Lucretius many other hypothesis and theories have existed suggesting species changing or adapting, in order to survive a changing world. Between six hundred and eleven and five hundred and thirty five BC, the Greek philosopher Anaxiamander was the first to hypothesise the common ancestor of man emerging from the worlds oceans, and adapting to life on land. One of these hypothesis came from Charles Darwin's own Grandfather, Erasmus Darwin. The British physician and poet proposed that life had changed over time. By the early 1800's a belief in evolution had been widely accepted by the British population.

The Beginning For Mammals.

Questionably, the evolution of man began during the Cretacious Period. When the first Mammals appeared upon the landscape, as the mighty dinosaurs were slowly becoming extinct. Before the great extinction event that saw the final demise of all non-avian dinosaur.

13 Evolution: Work In Progress

The evolutionary advance of man has the great extinction event in the form of six to twelve mile in diametre Asteroid that hit the Earth, creating the one hundred and thirty mile wide Chicxulub Crater in the Mexican Yucatan peninsula, sixty-five million years ago to thank. Before the great extinction event, mammals existed as small creatures the size of mice with some as large as modern day domestic cats.

The first mammals would scurry around at night for protection from the large dinosaurs. Burrowing under logs and rocks for safety during the day and warmth during the night. Their small size was advantageous as protection also from the nightmarish, unimaginable aftermath of the devastation of the asteroid collision, that marked the end of the dinosaur era on Earth.

Now with the Dinosaur wiped off the face of the planet, mammals have the freedom to expand their activity. Finally being able to come out of their little environments of safety, away from the gigantic, menacing creatures that held them into their nocturnal lifestyle. Food was not scarce immediately after the hellish firestorm and dense dust cloud that swept the surface of the planet when it must have seemed the end of the world, when the asteroid struck the Yucatan Peninsula. As they emerged from the burrows, gigantic, dead creatures littered the landscape and edible vegetation was plenty lying on the ground for such small creatures that had a big future on Earth.

As the Earth settled, becoming more habitable, the landmass was slowly moving into the position we see today with the subduction zones under the oceans continuously re-arranging the world. Uplifting mountainous areas and creating desert areas and areas of land, ideal for habitation, sustaining the continued diversification from the little furry mammals that were once frighteningly scared of those gigantic creatures that once ruled the Earth. Becoming larger and more adept at climbing for that food that was becoming harder to reach as the vegetation on Earth rejuvenated and adapted to the new moderate climate they were becoming climatised to, due to the movements of land, creating the oceanic currents and mountains with help from the polar regions which were forming, regulating the global average temperatures. Forests were dominating the land where the once little furry mammals habited. Lush forested areas were producing enough food and providing enough moisture to sustain the growth these creatures were experiencing.

Conditions on Earth was becoming ideal for the continual diversification and advance of the creature that would ultimately become ape, every other mammal on Earth today and ultimately, human.

Human Origins.

Humans did not evolve from apes, or any other primate known in the world today. That is a point that needs clarification. No evolutionary biologist, scientist or even Charles Darwin ever said they did. Man evolving from apes was something invented by the religious groups who oppose the evolution concept and refuse to accept evolution as the accepted, standard model that is recognised on how all life has diversified during a very long, slow process over millions of years. Originating from a common ancestor.

Evolution is the process that favours any genetic change that benefits changing environments. So, changes to DNA or genes cannot be major rapid changes. Genes are the very foundation of who we are as a species, as a living organism. Such changes can only be a slow, gradual process, taking hundreds of years for the slightest, smallest change to occur.

Take for instance, the very first, true human example found so far. Australopithecus afarensis, generally refered to as Lucy. Lucy is over three million years old. That's three hundred thousand generations from her and you and me. Three hundred thousand generations of extremely small changes between her, three million years ago and us in the modern world. Three hundred thousand generations is a very long time for the small genetic changes, or the evolutionary changes that have taken place between the first true human example and modern man.

With such small, slow, drawn out changes over such lengths of time. The general interpretation of a *Missing Link* is not going to happen. Such a *Missing Link*, is not going to be found, a *missing link* of skeletal remains of a half man, half ape is, well, impossible. Such a *missing link* does not follow the known laws of evolution.

Modern humans did evolve from a common ancestor of humans and apes, up to eight to nine million years ago. What were to become humans followed one evolutionary path, and what were to become the various ape species of today, followed their own evolutionary paths. That common ancestor has not been found yet, we have no idea what they looked like or how they lived. That elusive common ancestor is what is refered to as *The Missing Link*. It is not some half man, half monkey or half ape creature.

Chapter two

WHAT IS NOW

The evolution of modern human has certainly come a long way from our most ancient ancestor. That so-called missing link, that defined humans from other apes that exist on the Earth today.

Over nine million years of changes, adaptations so minute to notice over a short time of just a few centuries. Evolutionary changes are genetic changes. Genetic changes cannot be detected over a generation. Such changes are mutations. Mutations are not evolutionary changes, as they are more than likely not beneficial to future generations and will most likely die out with the next generation or certainly within a few generations. Evolution depends on changes that are permanent. Changes that will benefit future generations through continuous changes the planet undergoes, and changing environments.

The Earth does not stay the same, neither do living species of animal and fauna. The Earth is continually changing through subduction activity that is continually moving and re-arranging the landmass. Because of this movement and re-arranging, environmental changes are also occurring.

Over the last few hundred million years, during the reign of the dinosaur, the Triassic, Jurassic period, and the Cretaceous period. The period in which concluded with a massive meteorite slamming into the planet, seeing the final demise of the giant beasts. The Earth has undergone massive changes. Some were as a result of the massive meteorite impact. But, mainly these were due to the continents breaking up and forming the landmass we know on the planet today.

The subduction zones were continuously moving the Earths crust, forming the oceans and seas. Moving great landmasses into each other and forming the great mountainous areas we see all over the world today. These changes and cataclysmic events, created the global ocean currents and re-arranged the global weather patterns.

Life had to adapt and change with these changes or simply die out. A known law of evolution is to, adapt or become extinct. As with most natural laws, the known laws of evolution can be very unforgiving and decisive.

Since migrating out of Africa, The human species has undergone changes. Not as a whole, but during individual groups as they slowly migrated to different areas of the globe.

Those now living in desert areas and including Australia, have adapted to heat and dry conditions that those living in other areas, such as those living in the frozen tundra of Canada could not easily adjust to within any single generation. But, non-the-less they have adapted to opposite extremes of human abilities and tolerances, allowing them to live in relative comfort apart from each other, and still holding the individualistic ability to survive in this immense diversity of habitats.

When most people think of evolution in the Darwinian perspective, they think of the human body changing. Hopefully along with the known laws of evolution, that change is gradual, over an extremely long period of time. Not in the simplistic perspective that actually gives rise to unknown laws and impossibilities of any single generation, adapting to their environments or even multiple generations. The smallest evolutionary change will take tens of thousands of years, to millions of years. One must realize evolution is referring to the gene pool of any species. Not mutations or feature that can appear in any single individual or generation.

Other forms of evolution that have assisted various groups of humans to occupy various regions of the planet is: – Cultural Evolution and Social Evolution.

Cultural Evolution

Evolution, in the commonly used term, or, Genetic Evolution explains how life has diversified and changed through genetic changes over an extended time. From common ancestors, leading up to today. Cultural Evolution explains the social and cultural changes over an extended time. Changes that have benefitted cultural groups in a multi-linear sense that describes behaviour in separate cultures and societies.

Not only have the evolutionary changes been evident among human races of today. The cultural changes have also been evident, as human cultures have become more civilised, over an extended time. Civilisation through coexistence with one another through societies and the expectations placed upon individuals, in order to successfully coexist. Including the characteristics, behaviours, artefacts and ideas of individual groups and societies in their locations and sheer size of communities.

Cultural evolution would have begun as our earlier ancestors migrated out of Africa and settled in different areas and regions of the globe. With some instances being as the result of war and necessity for survival as climates and availability of food became scarce. Presenting the necessity to relocate to different areas, before the advent of farming about ten thousand years ago.

Evidence exists that cultural evolution has major factors, such as Geographical locations and also cultural diffusion, or the mixing of other cultures to help form a single culture belonging to other groups in other geographical locations.

A broad interpretation of Cultural Evolution is the ability to learn from one another. How some groups of human may be advanced further than some other groups belonging to different geographical locations, but the learned abilities benefitting these communities non-the-less.

Darwin's theory of evolution, by means of natural selection explains the adaptations by vertical transmission, the ability to receive genetic information from our parents. The ability to learn from our parents, authority figures and peers and so forth, is known as Oblique Transmission. Cultural Evolution has played a very important role in our evolution over millions of years, since our common ancestor with the apes.

Culture can never be underestimated. It is this that determines how we interact with others in our communities, in language with family members, friends and spouses or partners. Annual holidays, family traditions and religious

observances and customs are dictated by the cultural beliefs and traditions of any community.

In the modern world of multiculturalism, we often talk about the Maori culture, American culture, Chinese culture, Indian culture and so forth. These are usually observed from the native tribes or communities of different, various countries and people. And adopted by people from other cultures migrating to other areas.

Throughout human evolution, humans have been using their learned knowledge in making and refining their tools and weapons for hunting, fishing, gathering, cooking and eating. Over the last ten to thirty thousand years, the tools and implements have been created for farming and building shelters as communities, villages and cities have grown to cater and provide for the growing communities.

Cultural differences are evident around the world as communities grew and developed in isolation with populations growing while carrying these cultural differences through generations. These features were taught and refined from generation to generation, while being improved as the years passed.

Cultural evolution was considered as – Directional. That is, as human populations grew. Cultures became more complex. But recent understanding is considering cultural evolution as more inline with biological evolution. That is biological changes seem to be dictating cultural changes, or, natural selection, of the Darwinian evolution

sense. We see this in today's world as multiculturalism is expanding as more people are migrating to different geo graphical areas with increased ease as compared to earlier times, with cultures mixing and blending.

Social Evolution

Commonly confused with many aspects of Cultural Evolution, Social Evolution is the study of distinguishing aspects of human social groups. Though in differing degrees, all human groups posses it. Cultural and social traits must have begun as simplistic forms becoming gradually more complex as our biological changes became more complex.

As suggested by its name, Social Evolution dictates our socialistic traits and behaviour among groups and societies, with varying degrees being apparent among communities and societies belonging to different Geographical areas.

In 2010, a founder of modern sociobiology, Harvard University biologist, Edward O. Wilson proposed a new theory of Social Evolution. Primarily using insects in his argument, then using animals and humans social behaviour, arguing hereditary traits, environmental stimuli and past experiences and free will, is an illusion. Wilson argued that animal social behaviour is the result of genetic rules, governed by evolutionary processes. The same kind roughly formulated by Charles Darwin, back in the 1850's.

Charles Darwin would have had no way of realising his theory of natural selection was going to be the very basis

of what we understand about Evolution, when he published his formula - *On The Origin Of Species: By Means Of Natural Selection* in November 1859. Now in the 21st century. The known laws of evolution stand firm against his formula.

Humans are a very social species. Our entire structure is based on our social habits

The family unit for humans has always been the spine of our social activities, a source of learning from one another and developing our social skills. Modern technology is pushing our social behaviour in new directions with the Internet and especially social media, pushing new boundaries.

A digital identity is becoming an eminent part of our social behaviour and cannot be ignored. Social media has radically changed our social practices.

Chapter three

WHAT IS TO COME

Since I began writing with my first eBook, A History Of Life On Earth, then the follow up title, Distinctly Human: An Evolutionary Journey. Conversations I have had with various people have been interesting. The most interesting have been with Christians and other people of faith. Probably the biggest common denominator with these conversations about their understanding would be, many hold the strange, yet quaint belief that evolution does not explain adequately, the origins of life. These people seem a bit put off and perplexed when I inform them that evolution does not attempt to have an answer to how life began or originated.

The majority of these people, who I must admit are on social media, are creationists and intelligent design advocates. To me, they are trying to understand something they are not capable of comprehending. To them - God did it, is all they can comprehend, as limiting yourself to such simplicity prohibits the need to want to learn and ask questions. It is only through asking such questions we can expand our knowledge through learning.

God did it, seems the limit to learning for creationists and intelligent designers. Such simplicity eliminates any existence of complexities in their world. Don't get me wrong here, I am not saying these people do not understand evolution completely. They can seem quite knowledgeable and learned in the subject, but they tend to fall back on terms that are just not known in science.

For instance, they tend to refer to evolution as – Darwinism. Charles Darwin was not the first to realize life changed throughout time. But Darwin was the first to formulate a theory of Natural Selection. Which has become the standard theory on evolution that is understood around the world.

Another non-scientific belief favoured by creationists and intelligent designers is – six million years is not enough time for a species to evolve from its primate origins to modern man. Time scales seem to be incomprehensible to these people. Six million years is one heck of a long time for even the smallest adaptations to occur that have made modern humans into who we are today.

To put this timescale into some form of perspective - over the last one million years, since the very first true human ancestor – Lucy. There has been well over three hundred thousand generations. Three hundred thousand generations is a very long time for the tiniest, minutest adapta tions to our gene pool that have occurred from our first true human ancestor up to us. Modern humans.

You can only understand what you are capable of comprehending. These people who hold such firm beliefs that only god could create everything, while totally ignoring modern science and the scientific findings on how life originated through Abiogenesis and symbiogenesis. Then the changes or adaptations to changing environments through evolution, is rather sad and unfortunate for any human of today.

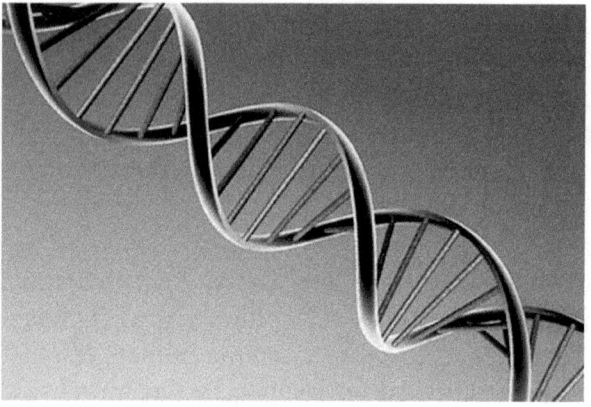

Abiogenesis is the theory on how life can arise under the proper conditions naturally from non living matter Symbiogenesis, or endosymbiotic theory - being a theory on how the origins of single celled organisms or Eukaryotes evolved into multi-celled organisms or Prokaryotes, in order to survive. And, how these multi-cells became dependant on each other to survive.

All species of animals and land plants are multi-cellular, as are some modern fungi and many algae

Evolution assumes life already exists before it can occur. Life must first exist before evolution can perform its wonders and beauty by adaptation. As previously mentioned. Evolution is a long, slow process, changes that occur to living forms is a reaction to changing environments and atmosphere. As the world continuously changes, so to, must life. Any failure to adapt can and will result in extinction of that species.

The purpose of evolution by natural selection is not to make any species perfect. The main purpose of natural selection is to ensure the survival and fitness of species to avoid extinction.

Fitness does not relate to the physical fitness of an individual. It is related to the fitness or best suited genes that are passed onto future generations.

Individuals belonging to a species, is not likely to survive and reproduce or breed if they are not properly adapted to their environments and surroundings. This means their genes are less likely to be passed onto the next generations, whereas, species that are properly adapted are going to be more settled, thus more likely to breed.

Variations are driven by many factors, including the environments and genes. But, it is only variations caused by the genes that are passed on.

Most forms of life that have ever existed on Earth are now extinct. The fossil record is a treasure trove of knowledge for which cannot be ignored. It is only through the fossil record we can analyse evolution through various stages of existence and changes that life has undergone throughout Earths history.

The facts and theory of evolution can take several meanings to different people who consider them. Evolution means small changes over a very long length of time. Referring to observed changes in organisms to their common ancestors through gene frequency.

Charles Darwin's theory of Natural Selection explains the mechanism of evolution.

Fact is something that is repeatedly observed and can be agreed on by peers. It can relate to findings and discoveries that no one can disagree with. It can also relate to well substantiated explanations.

When a scientist is explaining a theory such as Darwin's theory of natural selection, he is referring to the overwhelming evidence that supports the theory. The overwhelming evidence, that makes the theory – Fact.

Evolution by natural selection is so close to scientific fact that, no future findings or discoveries can ever significantly change or alter the existing theory as it stands today. So, evolution of natural selection, In the Darwin sense is regarded as – Fact.

31 Evolution: Work In Progress

Many of today's scientists and biologists believe the term – Theory is no longer appropriate or useful now when referring to the weight of existing evidence supporting Charles Darwin's theory of natural selection:-

- Richard Dawkins says - "*One thing all real scientists agree upon is the fact of evolution itself. It is a fact that we are cousins of gorillas, kangaroos, starfish, and bacteria. Evolution is as much a fact as the heat of the sun. It is not a theory, and for pity's sake, let's stop confusing the philosophically naive by calling it so. Evolution is a fact.*"

- Richard Lewontin wrote - "*It is time for students of the evolutionary process, especially those who have been misquoted and used by the creationists, to state clearly that evolution is fact, not theory*"

- Neil Campbell wrote in his 1990 biology textbook - "*Today, nearly all biologists acknowledge that evolution is a fact. The term theory is no longer appropriate except when referring to the various models that attempt to explain how life evolves... it is important to understand that the current questions about how life evolves in no way implies any disagreement over the fact of evolution*".

- Douglas J. Futuyma writes in *Evolutionary Biology* (1998) - "*The statement that organisms have descended with modifications from common ancestors—the historical reality of evolution—is not*

> *a theory. It is a fact, as fully as the fact of the earth's revolution about the sun"*.

Gradual Changes.

As previously mentioned and commonly known, Evolution is slow gradual changes over a very long period of time, changes that benefit future generations.
We see these gradual changes on ourselves. Slow changes, but certainly gradual as we move away from our ancient habits and other ways of life.

Before our ancestors began cooking their food and preparing with seasonings. Our third molars or, Wisdom teeth were extremely helpful to chew the tough tubers, nuts and raw meat that were the basic foods. Now, our third molars are not needed. But remain in the majority of humans. Evolution is slowly doing its job with our wisdom teeth, as today thirty percent of populations do not grow wisdom teeth, indicating they are slowly but surely being eradicated from their biological existence.

The same is true with our Appendix, very useful for our ancient ancestors in aiding their digestion of tough, and hard to digest food, but totally useless in modern humans. Our Appendix is looking as if it will remain for a very long time yet, doing absolutely nothing.

The Coccyx, or Tailbone is another example of a very useful body part, the remnant of a tail, when we shared part of our history with primates and the early mammals. It would be unfair for me to say the tailbone is now use less. As our Coccyx, does allow us to stand upright and keep our balance. The same role as a tail has on other primate today.

Not all adaptations are beneficial to species. Major defining advances for humans were our ability to stand upright, or Bipedalism. This gave our ancestors a great advantage above other species and predators. As bipedalism allowed our ancestors to stand taller above landmarks, and detect any present danger from potential predators, giving them the benefit to grab taller branches of trees.

The larger brain size of humans, present a problem for humans. The larger brain size requires a larger head to encase it. The head size compared to body size is distinctly human. No other species on earth has a similar head size to body size ratio as humans. This presents problems with birthing of human babies.

Chapter Four

ARE WE STILL EVOLVING

A big question that is fascinating after everything the human species has been through since our common ancestors with today's apes and all the changes, adaptation and evolutionary battles our ancestors fought to get to the stage we are at now is:- Are we still evolving? We know that evolution has made us who we are. But, are we still evolving?

That is a question scientists all over the world are trying to answer. Is the process that made us, now stopped? Or, is that process still continuing? As a species, we have adapted to changing environments. Many extreme environmental changes have forced us to change at a faster rate. Some slower rates have kept our evolution at a constant rate throughout our history from pre-human to human.

Evolution: Work In Progress

About three and a half billion years ago, life began. Most of the first life were, single celled organisms, some bacteria. Animals appear on the face of earth about six hundred million years ago. Then about seven million years ago, our first ancestors, the first hominins appeared. Then about two hundred thousand years ago, we, the human species first appeared. On an evolutionary time scale, humans have only just appeared among all the other species and life forms that have ever existed on Earth. Two hundred thousand years, is just the blink of an eye, compared to the three and a half billion years, life has been evolving on the planet.

The human species emerged on Earth about two hundred thousand years ago. About sixty thousand years ago, we emerged out of Africa and spread to every part of the planet. On the way, something happened to us that could suggest, the normal or usual laws of evolution may no longer be applicable to us. Tens of thousands of years ago, humans began protecting themselves from the environment and living in communities. Something no other species has ever done before. They began to make life easier by inventing things, such as tools, shelters and other objects that has definitely helped to make us who we are. Our ancestors made themselves clothes and fires to keep them warm, while other species developed coats of fur and some other species such as polar bears developed thick coats and blabber to keep them warm.

Humans are definitely the result of natural selection. But, thousands of years ago, we began making barriers that protected us from the harsh realities of the environment and other dangers that pose a threat to other species, with the use of simple technologies, that do not exist anywhere else in the natural world.

Have these barriers and buffers we have made for ourselves, stopped us from evolving further and some have excluded us from natural selection? That is a question that scientists have wondered since Charles Darwin's time. As a species, are we the same people that emerged from Africa two hundred thousand years ago?

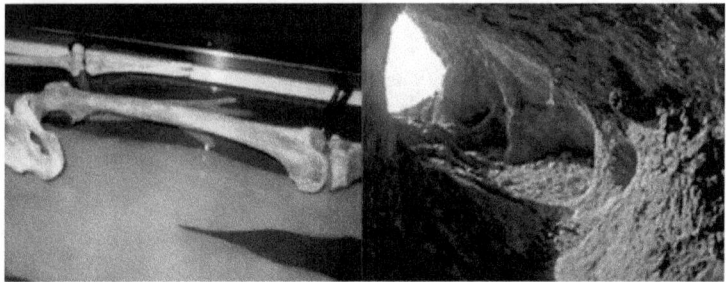

Left: The Red Lady Of Paviland. Right: Goats Hole Cave in Wales.

Contained in the Oxford University Natural History Museum in Oxford, England are the oldest bones of a modern human ever found in the United Kingdom, discovered in a Goats Hole cave in Wales by Rev. William Buckland, a Theologian, Geologist and Paleontologist, in 1823. The bones discovered by Buckland have a reddish oak pigmentation on them. Buckland first thought they were from the Roman era, and called them 'The Red Lady Of Paviland". But radio-carbon dating puts them at thirty three thousand years of age. Before the peak of the last ice-age. The remains are in fact, male of approxi mately 21 years of age and in remarkably good health at the time of death, thirty thousand years ago.

When this man was alive, he shared the planet with big wooly Mammoths that roamed the wilderness and Neanderthal. British physical anthropologist Dr. Alice Roberts says *"If I didn't know how old these bones were, that they had been radio-carbon dated to thirty three thousand years ago, I would believe you if you told me they were only a few hundred years old. Of course there's variations in skeletons, there's variation in our bodies, each of us will have a different skeleton. But, these bones fit within that range of variation."* Suggesting, there is nothing in that skeleton that tells us, we have changed within millenia. So, possibly, our use of invention and technology has excluded us from natural selection.

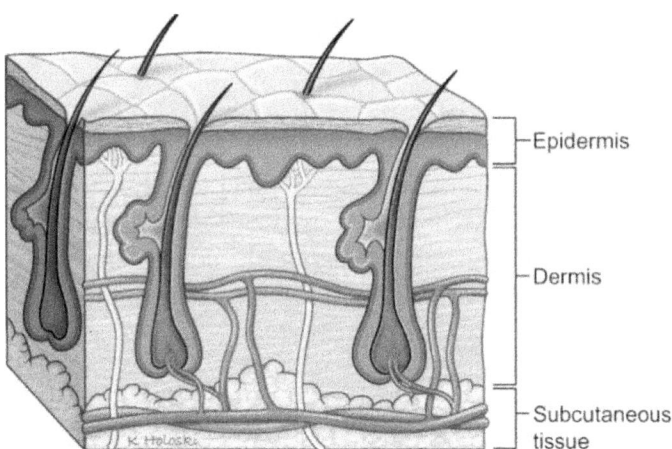

But then, if that were true, we would all be the same colour with the same hair colour. So that cannot be true that we have stopped evolving. But, it could also be a sign of how we have evolved separately, with our own cultures and traditions. And of course, that is only superficial. So, fits with in the range of regional adaptations and variations.

The colour of an individuals skin is determined by the genetic makeup of their biological parents. The single most important factor in determining your skin colour, is the melanin, which determines the colour pigmentation of your skin. Produced in skin cells called the Melanocytes, the colour is determined by tissue under the dermis and the hemoglobin circulating in the veins of the dermis.

As our ancestors spread out from Africa, to our own area of the world. The ultra-violet radiation penetrating the skin and controlling the biochemical reaction in the skin, determined the skin colour, from dark brown to lighter hues of skin pigmentation of individuals, through a process of natural selection.

Regional adaptations and variations are evident in civilizations, from various extreme geological locations around the world. For example, civilizations belonging to arctic regions and those from higher altitudes such as the higher Himilayas, have adapted differently from those closer to tropical and sub-tropical areas.

The Himilayan Sherpa's have evolved differently from the rest of us to adapt to the higher extremes, by requiring less hemoglobin in the blood that helps them to survive in less oxygenated air. The air in the higher Mountainous ranges are forty percent lower than that closer to sea level. The lower hemoglobin allows them to survive without the altitude sickness that will affect the rest of us, wider blood vessels allows them to receive enough oxygen without the fatal risk of higher hemoglobin levels. Adaptations that, are far from just superficial.

In some extreme environments at least, some civilisations did not stop evolving. Which raises questions about the rest of us. Genome research is allowing us to look deeper into human evolution than ever before. Allowing science to map our history. Genome research was a major breakthrough that many hoped would change the future of medicine. It has also opened a window on our past, which nobody anticipated. As, our genome contains a wealth of information about our history.

Pardis Sabeti an American computational biologist, medical geneticist and evolutionary geneticist at the Broad Institute of MIT and Harvard, which is often referred to as the Broad Institute, in Cambridge, Massachusetts, United States says, *"We are living records of our past. So, we can look at individuals from today and get a sense of how they all came to be this way."*

Essentially studying the DNA from thousands of people, it is possible to find genetic mutations that have become common in the last few thousand years. Or, looking for that adaptation that was so beneficial to future generations, it did not get lost in our history and spread throughout our populations. Such scientists have found over two hundred and fifty new genomes or recent genomes that are under natural selection. And can look at those changes and get a wider view or a global view at human evolution.

With Genome research and analysis, it is now clear we have evolved far more over the last two hundred thousand years, than anyone had previously imagined or anticipated.

Genetic research has opened up a lot of information about our past, that was unimaginable to our recent ancestors understanding. The recent history of us is written into our genes. But, our genetic history contains a lot of surprises.

A great defining event in our history was about ten thousand years ago, when our ancestors began farming. Farming has allowed for even the weakest in our populations to survive and allowed everyone to be sufficiently fed, through a stable food supply. Farming would most definitely have changed the way we evolved.

Farming presents its own problems. Crops have to be selected to suit various areas and climates. Crops can only be harvested during growth seasons and some growing seasons are going to fail due to climatic and weather conditions and some are going to flourish. European countries developed a staple that subsidized across all growing season. That staple was dairy.

Milk is full of energy, its very nutrient dense so it has lots of various vitamins and calcium. Milk has sugars in it, called lactose. A tolerance to that lactose was an evolutionary advantage that the winners will be better off.

Human babies produce an enzyme that breaks down that lactose, that enzyme is called lactase. But, human adults do not naturally produce that lactase, so an intolerance is a side effect. It was hugely advantageous for populations to develop that tolerance to such an important component to their diet. Some members of the population do not produce that tolerance and so are lactose intolerant. Lactose intolerance in healthy adults is more a discomfort that serious affliction, as this intolerance can produce lots if hydrogen, so produces flatulence in humans and in some cases diarrhea.

Before humans started farming, every human after the weaning period would have been lactose intolerant. But, with that need to have a component in their diet, those who developed that tolerance were the evolutionary winners.

The statistics for lactose tolerance is predominantly common in Europe and North-western Europe, also other areas where dairy farming is commonly a way of life. But is very rare in South-east Asia and other areas where dairy farming is not so common.

Farming has had a tremendous effect on our evolution over the last ten thousand years. Farming has actually been quite a main driver of our evolution. It seems the changes that we made to our world and history, has had a dramatic affect on our evolutionary changes than nature, itself, could throw at us.

Could we still be evolving today? Our way of life, our activities, hygiene and health has changed over the last few hundred years. Today in the modern world, almost anyone, irrelevant of genetic makeup and culture can survive long enough to pass on their genes, with a seemingly endless supply of food, medicine and sanitation has ensured fundamental changes have occurred within our species.

With the invention and implementation of these modern methods, life has certainly improved within the Twentieth century and twenty-first century as opposed to earlier patterns of life and death.

A few sobering statistics: during William Shakespeare's time (1564-1616) about 1 in 3 children died before they turned twenty-one years of age. By Darwin's birth, at the beginning of the nineteenth century (1809-1882), one in two children died before they turned twenty-one years of age. Simply because of their genetic makeup, or the genes they possessed, such as asthma and other conditions that were passed down from generation to generation. Many conditions that are now easily controlled or diseases, such as Cholera, for which people died in the millions, that have now been eliminated with modern medical procedures and medications. Now, in the twenty-first century, ninety-nine percent of modern babies survive long enough to pass on their genes.

At the time Charles Darwin formulated his theory on natural selection, a focus was on the mortality rate of populations. It was not until a half-century later, the focus was placed on the birth rate of populations, or fertility patterns, as the method of which genetic changes occurred in populations. Now, looking at the birth rate may seem the obvious focal point for natural selection. But during the middle of the twentieth century, that was quite a massive change of thinking and quite revolutionary. It changed the entire way scientists looked at evolution and in particular, natural selection.

Reproductive success is the main driving force behind natural selection. Reproductive success is what decides genetic frequencies.

Chapter Five

WHAT IS THE FUTURE FOR EVOLUTION

All life we see on Earth today, came to be through the processes of natural selection, from the very first living organisms, three point seven billion years ago. The very first simple organisms gradually became increasingly more complex, as time and the conditions of the planet changed. Common ancestors emerged and adapted to the changing conditions Earth presented them. The evolutionary process was completely dependant on those changes. If conditions were different, the evolutionary process of natural selection would have been different.

If the conditions changing environments presented our ancestors were different than those known to us, life may very well be totally different than today.

Lamarck

By the nineteenth century, and certainly by Charles Darwin's time, varying theories of evolution existed. In fact, the concept of evolution was well established to a point, different evolutionary explanations were a common talking point over the dinner table of Victorian England and many parts of Europe, as different theories saw the light of day, then died out as these theories did not stand the scrutiny of previous discoveries and findings. One of these theories was actually developed by Charles Darwin's grandfather, Erasmus Darwin.

In 1801, a French scientist, Jean-Baptiste Lamarck was developing his own theory of evolution.

Lamarck's theory followed two main points. The first being:-

- The Law of use and Disuse. And,
- The Law of Inheritance Of acquired Characteristics.

Diagram explaining Lamarcks theory that giraffes evolved their long necks by stretching to reach available food. These adaptations were inherited by their offspring.

Lamarck's theory stated: Characteristics, which are used more and more by an organism become stronger and bigger. Those that are not used eventually disappear. And any improved organism is passed on to its offspring.

His theory of inheritance of acquired characteristics is reliant on species changing and adapting on what they want and need. And, any changes are passed onto their offspring. For instance, Lamarck believed Elephants originally had short trunks. But, as food became scarce at lower levels, they stretched their trunks to reach the available food. These changes were inherited by their offspring until the trunk continued to increase in length. Lamarck believed the same occurred with Giraffes, who stretched their necks from need of food from a higher source, until their necks grew in length to what we now know of Giraffes.

According to Lamarck, the human body parts that are now not used, such as the Appendix and small toe will eventually evolve out of existence, as the adaptations over time will also be inherited by their offspring. This theory can also bear true to the wisdom teeth, which I pointed out earlier that are now evolving out of existence.

Lamarck's theory holds some credibility. Until you consider the continued existence of simple organisms. Lamarck's theory suggests, if true, simple organisms will evolve out of existence. His theory has no explanation for the continued existence of these simple organisms in the modern world. According to Lamark's theory, they should have evolved out of existence by now.

Cue: Charles Darwin's theory of natural selection, which has an explanation on the continued existence of simple organisms.

Misconceptions of Natural Selection

Charles Darwin's theory of natural selection is very easy to understand. But, it is commonly misunderstood among groups and individuals for variable reasons. As I previously mentioned in Chapter 3:- *You can only understand what you are capable of comprehending.* Some groups and individuals misinterpret natural selection through their own pre-conceived notion that - God did it, and generally, not understanding the concept of natural selection.

Natural selection is an important component of evolution as it explains the adaptations of an organism to its environment, whereas, evolution refers to the genetic changes in a population.

Other components of traits that make the general understanding of evolution is:

- Mutation
- Migration. And,
- Genetic Drift

All components are just as important as each other and none have any over-powering role from the rest.

An amusing misconception of natural selection is: natural selection somehow allowed the accidental origins of life. Or, somehow, natural selection initiated the first life. Evolution cannot occur if life does not exist. Evolution, especially natural selection relies on life already existing before it can occur. Before life can adapt to environmental changes, it must exist. Therefore evolution cannot have an answer to how life originated in the first place. But, this misconception persists.

Another persistent misunderstanding of natural selection is relating to death rates among populations. Natural selection is actually relating to the birth rate, as individuals who are more capable of reproducing will result in larger, healthier populations. Larger, healthier populations are stronger populations. An individual who lives a long life will result in more offspring than an individual who lives a shorter life. As previously mentioned, populations who are not properly adapted to their environments are less likely to reproduce than a population that is well adapted to their environment.

You can only understand what you are capable of comprehending

Natural selection is driven by - the survival of the fittest. This is an interesting misconception, as it seems to refer to the fitter individuals being able to survive in their environments against predators and fighting for food supplies as the main mechanism to survival of that individual. As if the fitter, stronger, faster individual has an evolutionary advantage over weaker, slower individuals. Evolution and especially natural selection favour stronger, fitter genes. Not the personal physical fitness of individuals. Physical fitness of an individual cannot be passed onto future generations. But, stronger genes will definitely be passed on. Any such genes must be beneficial to future generations. This misconception seems to be re-enforcing the misunderstanding that natural selection relates to the death rate of individuals and ignoring the main mechanism for natural selection is the birth rate.

The main driving force behind natural selection is the birth rate. Throughout human history, at times of conflict, natural disaster and times of immense uncertainty, higher birth rates have ensured the continuation of the pre-human and human populations. We see the same phenomena in the modern world. Poorer countries and groups experience high birth rates, in the hope some individuals survive into adulthood and carry on to reproduce. Thus, continuing the population. Natural selection will always favour those who maximize their reproductive success. Misunderstanding around this come from those who claim: "they should only have children if they can afford them", or other such statement. Such suggestions go totally against natural laws of species and especially natural selection, as if they did not reproduce at a higher rate, their populations will surely die out.

The Darwin Awards is a tongue in cheek honour of people removing themselves from the gene pool, thus, improving evolution. Unfortunately, the Darwin Awards is taken quite seriously as an accurate interpretation on natural selection on social media and email in the modern world. Natural selection actually refers to favourable traits being passed onto future generations and less favourable adaptations or traits will die off and eventually be removed from the gene pool. Evolution and natural selection is not interested in individuals, who do stupid things that result in their death.

Evolution is not a safety mechanism against accidents and stupidity. Humans are not immune to accidents. Evolution is not interested in making any individual or any particular species perfect. The main purpose of evolution is to ensure species avoid extinction. No more, no less.

Apparently, Charles Darwin invented natural selection. Not only is that a ridiculous statement, it is obviously borne from a deep ignorance of both Darwin and the entire concept of evolution. Did Einstein invent the increased relativistic mass (m) of a body coming from the energy of motion of the body - that is, its kinetic energy (E) - divided by the speed of light squared (c2). That equation expresses the fact that mass and energy are the same physical entity and can be changed into each other? Of cause not, the mass, kinetic energy and the speed of light already existed. Einstein simply discovered his theory of $E=mc^2$ Through experimentation and observations, as opposed to inventing it. Did Isaac Newton invent gravity? Not likely as gravity has been one of the forces of nature since our planet formed. Newton simply discovered it through experimentation and observations, when an apple fell off its tree.

Considering evolution had been occurring for billions of years before Darwin was born.

It is ludicrous to state Darwin invented natural selection. Charles Darwin was the first naturalist to formulate a theory of natural selection. Natural selection would have to exist before Darwin could formulate his theory, based on his observation and testing. At the same time Darwin was working on his theory, another naturalist – Alfred Russel Wallace was also working on the same theory, miles apart from one another. The first actual public explanation of natural selection was a joint presentation involving both Charles Darwin and Alfred Wallace. Darwin gets the credit for the theory of natural selection because, he was the first documented, to publish a book on the subject, in 1859.

Persistence in the misconception, evolution continues to evolve randomly, or by chance. Evolution by natural selection dictates adaptation within a species can only occur with specific variations that are possible with any given species. Or, in other words - Adaptations are reliant on the possibilities of changes with a species. So, according to the known laws of evolution and natural selection, adaptation can never be random or, as the result of chance. As, they are completely dependant on what is possible.

There should be evidence of an intermediary species in the fossil record. Or, there should be evidence of a half man/half ape species. Evolutionary change is based on changes to the genetic makeup of species and not on some fantasy notion of changes within a single genera tion. These genetic changes can take an extremely long time to gradually occur. Wouldn't it be great if we could adapt to resistance to many ailments or diseases prior to taking an overseas vacation? Evolution just does not work that way. Adaptation to our genetic makeup, are not superficial and take many generations for the smallest change to occur.

Did humans evolve from apes as the now infamous claim suggests? Of cause not, not even Charles Darwin said humans evolved from apes. But that is part of the persistence of the myth. Advocates of the myth claim Darwin mentioned it in his best selling publication – *On The Origin Of Species: By Means Of Natural Selection*.

That claim could be no further from the truth. Darwin only mentions humans or man once in his entire book. In the last chapter, third paragraph from the end, Darwin wrote:

In the distant future, light will be thrown on the origin of man and his history.

Man evolved from apes was a term invented by religious leaders and quickly adopted by other groups who refused to accept Darwin's theory on natural selection. Religious leaders know they can successfully make such outlandish claims, as they are fully aware their followers will not check the claim for themselves. So, such claims can gain traction that persists for decades, or, in this case, centuries.

Allegations Darwin had made such a claim in his 1859 publication, could have been the reason Darwin teasingly mentioned the same claim in his 1871 publication – *The Descent Of Man*. In making the comment, Darwin suggested man and apes share a common ancestor. In which one branch of a species followed their own evolutionary path to ultimately become modern man. The other branched off and followed their evolutionary path to ultimately become modern ape.

Evolution is just a theory. Yes, evolution is just a theory. But, it is the interpretation of the word – Theory, that is at question. The general definition of theory is different from the scientific definition of – theory. In science, theory is an explanation gained through the scientific method, with experimentation and continued testing or, confirmed through observation. Scientific theory is generally regarded as – Fact.

Today, evolution is so close to fact that no future findings or discoveries could significantly change or alter the original theory. Today, evolution is the standard model that is recognised on how all life has diversified or changed over the years, originating from a common ancestor. Diversifying into different lineages, then different species and advancing into the modern life we see all over the world today, through natural selection. Favouring any genetic changes that benefit future generations for changing environments.

Yes, evolution is just a theory… A scientific theory. Yes, evolution is – Fact.

The human species has been evolving for the last nine million years, since our common ancestor with the apes. There is no evidence that suggests we have stopped evolving. Natural selection says, if you stop evolving, you will become extinct.

As suggested, species that are not properly adapted to their environments will not breed successfully. The low birthrate of most modern societies are quite concerning, as each couple having one to two children are ultimately replacing themselves and not growing the populations of those societies. When the parents die, the number in the population is remaining the same and even decreasing if some offspring do not outlive their parents, and parents who only have one child.

Societies that are less modern are experiencing far more birthrates, suggesting they are far more adapted to their environment, whose populations are growing and diversifying. So the entire concept of evolution could be changing its preferences in the continuation of the human species over other groups of society.

It stands to reason, the societies that are not experiencing sustainable birthrates should welcome migrants from those societies that are experiencing higher birthrates, allowing these societies to grow and mix in a more sustainable fashion.

As the world gets increasingly smaller, as technology and the ease of travel from continent to continent and country to country get more complex, allowing us to move and migrate to different areas of the world with the seemingly lack of geographic and physical borders, that once presented major problems for our ancient and more modern ancestors. Evolution is most likely to take another path to our future.

As once isolated communities integrate with increasing ease among communities, which only a few decades ago, would have been impossible due to barriers prohibiting such movements, and financial constraints, as travel becomes more within reach of these communities. Future generations will find multiculturalism becoming more acceptable and affordable.

The Canadian Royal Commission on Bilingualism and Biculturalism, introduced, what is known as the modern political awareness of multiculturalism in 1971. The ideology spread to modern western English speaking countries and territories.

Australia, since slowly dissolving their White Australia policies after World War 2 has adopted the policies of multiculturalism and embraced to ideology in 1973, along with most European Union members. Although some European nations, notably, Netherland and Denmark have returned to a form of monoculturalism which originally came into existence during the eighteenth and nineteenth centuries, in an effort to protect cultures and traditions.

In the future world, with an increase in multiculturalism, hopefully racial intolerance, prejudice and bigotry will decrease among extreme rightwing members of these communities. Or else they face a real threat of extinction as they fail to accept changes, they may become improperly adapted to their ancestral homes and face the same fate of all other living form of life on Earth, that failed to adapt to their changing environments. Though the failure of these people to adapt may not be a bad thing for the future and stability of humanity. Racial and prejudiced attitudes may die off as communities become more multicultural.

Globalisation and capitalism are two main drivers of change the world is experiencing in the twenty first century, forcing more and more people to seek new lands to migrate into and spread their roots for future generations of family. Conflicts and unstable economies are projected to increase this flow of refugees seeking their freedoms and safety.

As the face of the worlds human populations change, so to will the future face of evolution by natural selection, driving that change.

Cultural indicators continually drive evolution around within short lengths of time as cultures change. Cultures are rapidly changing as multiculturalism increases worldwide. This could quite successfully be argued as a cultural explosion as cultures that have previously been isolated, are now increasingly free to mix.

Only a massive environmental change will send evolution onto a new path. These significant environmental changes will take natural selection on to a path that will ensure by triggering genetic changes in future populations.

Where evolution is taking us now, at this particular point is unknown. Only a rapid environmental shift will alter the future of our natural selection. Culture plays a major part and culture is exploding at this particular time.

Natural selection will continue to change with any significant shift with culture. Culture is a main part of the environment. We may not see this, as we are part way through it. But, over all, it will make up a significant change in centuries to come.

Such is with evolution, in particular natural selection. Changes are so slow and subtle, generations do not notice these changes. But over an extended time period, they are very noticeable as comparative to earlier generations.

Genetic Modification

Genetic modification, or, Genetic engineering has been a practice of humans for thousands of years. Selective ge netic modification is responsible for the many modern breeds of dog, Cats and many other domestic and agricultural breeds of animal alive and benefitting humans in many different ways.

Genetic modification is a practice in plants. Favouring specific plant DNA that benefit groups of humans in some acceptable types of plant and some that remain controversial.

The direct manipulation of the genome of organisms by the use of biotechnology, specifically intended to change the genetic makeup of cells, including the transfer of genes, within and across species boundaries to produce improved or novel organisms. New DNA is inserted into the host genome, by first identifying, isolating the genetic material. Then copying the genetic material to generate a new genetic sequence of the host. This can be achieved by recombining the genetic makeup of species or breeds.

Plants, animals or micro organisms that have changed through genetic engineering, are termed genetically modified organisms, or GMO.

The controversy surrounding these genetic alteration, is increasing as technology and the practice improves. Raising questions into its acceptance as genetic modification is increasing around the modification of DNA with humans.

The human embryo is the very beginning of human life. From the fourth day after conception, the human embryo is a multicellular diploid eukaryote, this is defined as the developing organism, until after the eighth week of fertilisation. Then the unborn baby is generally referred to as, the foetus.

The first test tube baby, as they were referred to was created in 1978 by Dr. Jeff Steinberg, at his clinic for in-vitro fertilisation in Los Angeles, California. Since then, he has become instrumental in the creation of many new babies.

Over recent decades, the creation of designer babies is undergoing perfection in technique. Such clinics as Dr Steinberg's are routinely screening embryos for genetic diseases. Some are beginning to offer prospective parents the option of cosmetic traits. Such as hair colour, eye colour and the sex of the child.

It is with such designer babies, the way the world is heading. As the technology and procedures improve. This will most definitely be the way of the future for those who wish to choose traits and the sex of their offspring.

An ability to manipulate our genes while selecting offspring with such procedures could severely alter the course of our evolution, in the search for physical perfection.

Chapter six

CONCLUSION

Its clearly obvious we have continued to evolve throughout our history. Just as clear as we are continuing to evolve, and will do into the future. The way we will evolve depends on how the world changes and how we change the world we live in.

Technology is advancing faster today, than at any other time in our history. Our world is changing faster than ever before. As for evolution, no one can accurately tell where the changing world is going to take us. As our world continues to change and technology improves, the only certainty is, we will either continue to evolve. And rapidly, or we will, simply out-do ourselves and stop evolving. If that occurs, we will die out. As a species, we will simply not exist.

61 Evolution: Work In Progress

Humans occupy a special place on Earth and its history. Our achievements and the tools we have created. The homes we have built ourselves, our communities, cultures, traditions that have resulted in humans occupying every area of the globe. The massive cities and inventions have ensured our place on the planet. But, humans are not too special. Ninety nine percent of species that have existed over the Earths history have become extinct and simply died off. Eventually, our species will join them.

Life is fragile. A global catastrophe could wipe us all out. But, even a small number of survivors could continue the evolutionary journey, if they adapt to the new world they inhabit. They could continue a journey that began three point seven billion years ago.

Humans have not always been human. Modern humans are very different from the early stone age humans. Modern man is very different from the human of the future. As the world rapidly changes, so to, will our evolution.

Over the last nine million years, since our earliest ancestor. Our evolution has made dramatic advancements. Changes that have not occurred in any other species. Our ancestors adapted to their changing, unique environments in a way that has never existed before. Our ancestors fought natural elements that should have been a boundary to their limits. But, they persevered, survived and adapted to those impossible odds. Modern humans owe their very existence to their ancestors and the natural battles they fought to allow our existence.

No human on Earth today have ancestors that did not adapt to their changing world. If they did not adapt, they would have died off. No human alive on Earth today had ancestors who did not survive into adulthood. We deserve that special place in Earth history.

But, where will our evolutionary journey take us? What is next for the human species?

One factor is a certainty. Evolution certainly is - A Work In Progress.

Appreciation

I would personally like to thank you for purchasing and reading this book. I enjoyed writing ***Evolution: Work In Progress***. I hope it was enjoyable for you to read.

Bruce Alpine.

For more Titles available from Bruce Alpine. Visit the Website:

https://brucealpine.com

Contact The Author:

bruce-a@brucealpine.com

EVOLUTION: Work In Progress

Bruce Alpine.

ePub ISBN: 978-0-9941053-7-0

Print ISBN: 978-1-5398826-3-3

www.ingramcontent.com/pod-product-compliance
Lightning Source LLC
Chambersburg PA
CBHW070134210526
45170CB00013B/1033